Bibliographic information published by the German National Library:

The German National Library lists this publication in the National Bibliography; detailed bibliographic data are available on the Internet at http://dnb.dnb.de .

Imprint:

Copyright © 2018 GRIN Verlag
Print and binding: Books on Demand GmbH, Norderstedt Germany
ISBN: 9783668851016

This book at GRIN:

https://www.grin.com/document/450780

Deapon Biswas

The Four Biswas Members

GRIN Verlag

GRIN - Your knowledge has value

Since its foundation in 1998, GRIN has specialized in publishing academic texts by students, college teachers and other academics as e-book and printed book. The website www.grin.com is an ideal platform for presenting term papers, final papers, scientific essays, dissertations and specialist books.

Visit us on the internet:

http://www.grin.com/

http://www.facebook.com/grincom

http://www.twitter.com/grin_com

THE FOUR BISWAS MEMBERS

DEAPON BISWAS

Transport Officer, Private Concern, Chittagong, Bangladesh.

Abstract

So far combinations and permutations are discussed with different theorems in algebra. In this paper I apply assembly analysis to get the theorems easy and memorable. After assembly analysis applied there becomes a lot of new theorems and all the theorems get a new face by summation methods. Formations mean the selections of a random experiment where order is not taken into account and repartitions are allowed. Homogenations mean the selections of a random experiment where order is taken into account and repartitions are allowed.

Keywords

Combinations, general combination theorem, combination distribution, permutations, general permutation theorem, permutation distribution, formations, general formation theorem, formation distribution, homogenations, equigenous expression, general homogenation theorem, homogenation distribution.

Article Outline

1. Introduction
2. Findings
    2.1 Combination space
    2.2 Combination event
    2.3 General combination theorem
    2.4 Combination distribution
    2.5 Permutation space
    2.6 Permutation event
    2.7 General permutation theorem
    2.8 Permutation distribution

2.9 Formation space

2.10 Formation event

2.11 General formation theorem

2.12 Formation distribution

2.13 Homogenation space

2.14 Homogenation event

2.15 multinomial expression

2.16 General homogenation theorem

2.17 Homogenation distribution

  3. Conclusions

1. Introduction

Formation is an outcome of a random experiment of M different components taken V at a time, the numbers M and V have one of the relations $M < V, M = V$ or $M > V$, where for combinations there exists $M \geq V$. It is main difference to combinations with formations is that in case of formations the so called M components designated by M sides of a die. In case of combinations the M components can not be designated by M sides of a die. Homogenation is an outcome of a random experiment of M different components taken V at a time, the numbers M and V have one of the relations $M < V, M = V$ or $M > V$ where for permutations there exists $M \geq V$. It is main difference to permutations with homogenations is that in case of homogenations the so called M components designated by M sides of a die. In case of permutations the M components cannot be designated by M sides of a die. Carefull attention must be paid that M and V always be positive integers. A general query informed you that of the terms "component" and "place". Usually we can say every component has its place in a formation where a formation is of V components recorded in a Q-tuple. As repetitions are allowed we should say every q^{th} place, $q = 1, 2, 3, \ldots, Q$ has the capacity to hold P_{tq} components to make a formation. In other words we can say the qth place can be filled with one of the M components repeating P_{tq} times. Before going to original discussion we recall the partition $P_t = \left(P_{t1} + P_{t2} + P_{t3} + \cdots + P_{tq} + \cdots + P_{tQ}\right)$ where the Q components indicates Q runs with associate lengths P_{tq} ; $q = 1, 2, 3, \ldots, Q$ or Q kinds with associate lots P_{tq} ; $q = 1, 2, 3, \ldots, Q$.

2. Findings

2.1 Combination space

A combination space is a set of all possible combinations (outcomes) of an experiment from a parent assembly A where the outcomes do not take order of the components into account. Let a combination space contains T possible outcomes then the combination space denoted by $C\begin{Bmatrix} A \\ V \end{Bmatrix}$ is

$$C\begin{Bmatrix} A \\ V \end{Bmatrix} = \{C_1, C_2, C_3, \cdots, C_t, \cdots, C_T\} \quad\text{————— (1)}$$

$$\text{where, } V = 1, 2, 3, \ldots, N$$
$$N = \text{Parent component number.}$$

Example 1: Set a combination space of the experiment "4 letters a, b, c, and d select 3 at a time ".
Solution: We have given $A = (a, b, c, d)$ and $V = 3$.
Thus the combination space is
$$C\begin{Bmatrix} (a, b, c, d) \\ 3 \end{Bmatrix} = \{(a, b, c), (a, b, d), (a, c, d), (b, c, d)\}.$$

Theorem 1: The number of combination of N different components taken V at a time denoted by $C\begin{pmatrix} N \\ V \end{pmatrix}$ is
$$C\begin{pmatrix} N \\ V \end{pmatrix} = \frac{(N-V+1)(N-V+2)(N-V+3)\ldots N}{V!} \; ; \; V \leq N \quad\text{————— (2)}$$
Using the summation method we get (2) as
$$C\begin{pmatrix} N \\ V \end{pmatrix} = \sum_{k_1=1}^{(N-V+1)} \sum_{k_2=k_1+1}^{(N-V+2)} \sum_{k_3=k_2+1}^{(N-V+3)} \cdots \sum_{k_v=k_{v-1}+1}^{N} C \quad\text{——— (3)}$$

2.2 Combination event

It is a special kind of subset of a combination space where the combination members are to be have same first components, same second

3

components, same third components and so on same v^{th} components. Suppose the combination members

$$C_1 = (C_{11}, C_{12}, C_{13}, \dots, C_{1v}, \dots, C_{1V})$$
$$C_2 = (C_{21}, C_{22}, C_{23}, \dots, C_{2v}, \dots, C_{2V})$$
$$C_3 = (C_{31}, C_{32}, C_{33}, \dots, C_{3v}, \dots, C_{3V})$$
$$\vdots$$
$$C_t = (C_{t1}, C_{t2}, C_{t3}, \dots, C_{tv}, \dots, C_{tV})$$

where, $C_{11} = C_{21} = C_{31} = \dots = C_{t1}$
$C_{12} = C_{22} = C_{32} = \dots = C_{t2}$
$C_{13} = C_{23} = C_{33} = \dots = C_{t3}$
$\vdots$
$C_{1v} = C_{2v} = C_{3v} = \dots = C_{tv}$

Then the combination event denoted by $C\left\{{A \atop V/^v A}\right\}$ consists of $C_1, C_2, C_3, \cdots, C_t$ i.e.,

$$C\left\{{A \atop V/^v A}\right\} = \{C_1, C_2, C_3, \cdots, C_t\} \underline{\hspace{3cm}} (4)$$

Here A is a parent assembly containing N components, V is the number of components occurred in a combination member and $^v A$ is an identified component assembly containing v identified components. The combination event contains any number of combination members and starts with any combination member of $C\left\{{A \atop V}\right\}$ contains the condition supports.

Example 2: Find the combination event of the example 1 where identified components are $^2 A = (a, b)$.

Solution: The combination event is

$$C\left\{{(a, b, c, d) \atop 3/(a, b)}\right\} = \{(a, b, c), (a, b, d)\}.$$

Theorem 2: The number of combination of N different components taken V at a time whose first v components are identified, denoted by $C\left({N \atop V/v}\right)$ is

$$C\left({N \atop V/^v A}\right) = \frac{(N-V+v-k_v+1)\,(N-V+v-k_v+2)\dots\dots(N-k_v)}{(V-v)!} \underline{\hspace{2cm}} (5)$$

Using the summation method we get (5) as

4

$$C\begin{pmatrix} N \\ V/^v A \end{pmatrix} = \sum_{k_{v+1}=k_v+1}^{(N-V+v+1)} \sum_{k_{v+2}=k_{v+1}+1}^{(N-V+v+2)} \sum_{k_{v+3}=k_{v+2}+1}^{(N-V+v+3)} \cdots \sum_{k_v=k_{v-1}+1}^{N} C$$

$$\rule{6cm}{0.4pt}$$ (6)

where C is a constant quantity taking unit value.

2.3 General combination theorem

Theorem 3: The number of combinations in a combination space of N different components taken V at a time where there occurred Y particular components taken from $U \leq V$ particular components (limited size of components) that occurred in the parent component assembly of N components denoted by $C\begin{pmatrix} N & U \\ V & Y \end{pmatrix}$ is

$$C\begin{pmatrix} N & U \\ V & Y \end{pmatrix} = C\begin{pmatrix} U \\ Y \end{pmatrix} C\begin{pmatrix} N-U \\ V-Y \end{pmatrix} \rule{3cm}{0.4pt}$$ (7)

$$\text{where, } N \geq V$$
$$U \leq V$$
$$U - N + V \leq Y \leq U$$

Y takes only non-negative values.

Proof: Let the parent component assembly is of N components. The combination members are of V components. Now we can select Y components out of U components without restrictions is $C\begin{pmatrix} U \\ Y \end{pmatrix}$. Then for each combination member there remains (V–Y) components to select out of (N–U) components. The number of these selections is $C\begin{pmatrix} N-U \\ V-Y \end{pmatrix}$. Hence finally we get the total number of combinations is $C\begin{pmatrix} U \\ Y \end{pmatrix} C\begin{pmatrix} N-U \\ V-Y \end{pmatrix}$.

Example 3: Let the parent component assembly M = (A, B, C, D, E). Find the number of combinations of the following and then write the combinations.

(i) $C\begin{pmatrix} 5 & 3 \\ 3 & 3 \end{pmatrix}$, (ii) $C\begin{pmatrix} 5 & 3 \\ 3 & 2 \end{pmatrix}$, (iii) $C\begin{pmatrix} 5 & 3 \\ 3 & 1 \end{pmatrix}$, (iv) $C\begin{pmatrix} 5 & 2 \\ 3 & 2 \end{pmatrix}$, (v) $C\begin{pmatrix} 5 & 2 \\ 3 & 1 \end{pmatrix}$.

For (i) to (iii) the components of limited size is (A, B, C)

For (iv) to (vi) the components of limited size is (A, B)

Solution:

(i) $C\begin{pmatrix} 5 & 3 \\ 3 & 3 \end{pmatrix} = C\begin{pmatrix} 3 \\ 3 \end{pmatrix} C\begin{pmatrix} 5-3 \\ 3-3 \end{pmatrix} = 1 \times 1 = 1.$

Now the combination is (A, B, C).

(ii) $C\begin{pmatrix} 5 & 3 \\ 3 & 2 \end{pmatrix} = C\begin{pmatrix} 3 \\ 2 \end{pmatrix} C\begin{pmatrix} 5-3 \\ 3-2 \end{pmatrix} = 3 \times 2 = 6.$

Now the combinations are (A, B, D), (A, B, E), (A, C, D), (A, C, E), (B, C, D), (B, C, E).

(iii) $C\begin{pmatrix} 5 & 3 \\ 3 & 1 \end{pmatrix} = C\begin{pmatrix} 3 \\ 1 \end{pmatrix} C\begin{pmatrix} 5-3 \\ 3-1 \end{pmatrix} = 3 \times 1 = 3.$

Now the combinations are (A, D, E), (B, D, E), (C, D, E).

(iv) $C\begin{pmatrix} 5 & 2 \\ 3 & 2 \end{pmatrix} = C\begin{pmatrix} 2 \\ 2 \end{pmatrix} C\begin{pmatrix} 5-2 \\ 3-2 \end{pmatrix} = 1 \times 3 = 3.$

Now the combinations are (A, B, C), (A, B, D), (A, B, E).

(v) $C\begin{pmatrix} 5 & 2 \\ 3 & 1 \end{pmatrix} = C\begin{pmatrix} 2 \\ 1 \end{pmatrix} C\begin{pmatrix} 5-2 \\ 3-1 \end{pmatrix} = 2 \times 3 = 6.$

Now the combinations are (A, C, D), (A, C, E), (A, D, E), (B, C, D), (B, C, E), (B, D, E).

2.4 Combination distribution

Theorem 4: A random variable Y is said to follow combination distribution if it assumes only non-negative values and its probability mass function is given by

$$P(Y) = C(Y; N, U, V) = \frac{C\begin{pmatrix} U \\ Y \end{pmatrix} C\begin{pmatrix} N-U \\ V-Y \end{pmatrix}}{C\begin{pmatrix} N \\ V \end{pmatrix}} \; ; \; U-N+V \leq Y \leq U$$

$$= 0 \; ; \text{otherwise} \qquad\qquad (8)$$

The three independent finite constants N, U and V are known as the parameters of this distribution. Combination distribution is a discrete distribution as Y can take only the non-negative values under the interval $U-N+V \leq Y \leq U$. Any variable which follows combination distribution is known as combination variate and denoted by the symbol $Y \sim C$ (N, U, V).

2.4.1 Moments

The first four moments about origin of combination distribution are

$$\mu_1' = E(Y) = \frac{UV}{N}.$$

$$\mu_2' = E(Y^2) = \frac{UV(U-1)(V-1)}{N(N-1)} + \frac{UV}{N}.$$

$$\mu_3' = E(Y^3) = \frac{UV(U-1)(U-2)(V-1)(V-2)}{N(N-1)(N-2)} + \frac{3UV(U-1)(V-1)}{N(N-1)} + \frac{UV}{N}.$$

$$\mu_4' = E(Y^4)$$

$$= \frac{UV(U-1)(U-2)(U-3)(V-1)(V-2)(V-3)}{N(N-1)(N-2)(N-3)} + \frac{6UV(U-1)(U-2)(V-1)(V-2)}{N(N-1)(N-2)}$$

$$+ \frac{7UV(U-1)(V-1)}{N(N-1)} + \frac{UV}{N}.$$

Now variance $\mu_2 = \mu_2' - \mu_1'^2$

$$= \frac{UV(U-1)(V-1)}{N(N-1)} + \frac{UV}{N} - \frac{U^2V^2}{N^2}$$

$$= \frac{UV(N-U)(N-V)}{N^2(N-1)}.$$

2.5 Permutation space

A permutation space is a set of all possible permutations (outcomes) of an experiment from a parent assembly A where the outcomes take order of the components into account. Let a permutation space contains T possible outcomes then the permutation space denoted by $P\{^A_V\}$ is

$$P\{^A_V\} = \{P_1, P_2, P_3, \cdots, P_t, \cdots, P_T\} \quad\underline{\hspace{3cm}} (9)$$

where, V is the number of components occurred in an outcome.

Example 4: Set a permutation space of the experiment "arrange 4 digits 1, 2, 3 and 4 taken 3 at a time".

Solution: We have given A = (1, 2, 3, 4) and V = 3.
Thus the permutation space is

$P\{^{(1,2,3,4)}_3\}$ = {(1, 2, 3), (1, 3, 2), (1, 2, 4), (1, 4, 2), (1, 3, 4), (1, 4, 3), (2, 3, 4), (2, 4, 3), (2, 3, 1), (2, 1, 3), (2, 4, 1), (2, 1, 4), (3, 4, 1), (3, 1, 4), (3, 4, 2), (3, 2, 4), (3, 1, 2), (3, 2, 1), (4, 1, 2), (4, 2, 1), (4, 1, 3), (4, 3, 1), (4, 2, 3), (4, 3, 2)}.

Theorem 5: The number of permutations of N different components taken V at a time denoted by $P\binom{N}{V}$ is

$$P\binom{N}{V} = N(N-1)(N-2)...(N-V+1) \qquad\qquad \text{————— (10)}$$

Using the summation method we get (10) as

$$P\binom{N}{V} = \sum_{k_1=1}^{N} \sum_{k_2=1}^{(N-1)} \sum_{k_3=1}^{(N-2)} .....\sum_{k_V=1}^{(N-V+1)} C \qquad\qquad \text{————— (11)}$$

2.6 Permutation event

It is a special kind of subset of a permutation space where the permutation members are to be have same first components, same second components, same third components and so on same v^{th} components. Suppose the permutation members are

$$P_1 = (P_{11}, P_{12}, P_{13}, ..., P_{1v}, ..., P_{1V})$$
$$P_2 = (P_{21}, P_{22}, P_{23}, ..., P_{2v}, ..., P_{2V})$$
$$P_3 = (P_{31}, P_{32}, P_{33}, ..., P_{3v}, ..., P_{3V})$$
$$\vdots$$
$$P_t = (P_{t1}, P_{t2}, P_{t3}, ..., P_{tv}, ..., P_{tV})$$
$$\text{where}, \quad P_{11} = P_{21} = P_{31} = ... = P_{t1}$$
$$P_{12} = P_{22} = P_{32} = ... = P_{t2}$$
$$P_{13} = P_{23} = P_{33} = ... = P_{t3}$$
$$\vdots$$
$$P_{1v} = P_{2v} = P_{3v} = ... = P_{tv}$$

Then the permutation event denoted by $P\left\{{A \atop V/{}^v A}\right\}$ consists of $P_1, P_2, P_3, ...,$ P_t i.e.,

$$P\left\{{A \atop V/{}^v A}\right\} = \{P_1, P_2, P_3, \cdots, P_t\} \qquad\qquad \text{————— (12)}$$

Here A is a parent assembly containing N components, V is the number of components occurred in a permutation member and ${}^v A$ is an identified component assembly containing v identified components.

Example 5: Find the permutation events of the example 4 where the identified components are (i) ${}^1 A = (2)$, and (ii) ${}^2 A = (3, 1)$.

Solution: The permutation events are

(i) $P\left\{{(1,2,3) \atop 3/(2)}\right\} = \{(2, 1, 3), (2, 3, 1)\}$

(ii) $P\left\{\dfrac{(1,2,3)}{3/(3,1)}\right\} = \{(3,1,2)\}$.

Theorem 6: The number of permutations of N different components taken V at a time, whose first v components are identified, denoted by $P\left(\dfrac{N}{V/^vA}\right)$ is

$$P\left(\dfrac{N}{V/^vA}\right) = (N-v)\,(N-v-1)\,(N-v-2)\,\ldots\,(N-V+1) \quad\text{------ (13)}$$

Using the summation method we get (13) as

$$P\left(\dfrac{N}{V/^vA}\right) = \Sigma_{k_{v+1}=1}^{(N-v)}\,\Sigma_{k_{v+2}=1}^{(N-v-1)}\,\Sigma_{k_{v+3}=1}^{(N-v-2)}\,\ldots\,\Sigma_{k_V=1}^{(N-V+1)}\,C \quad\text{------ (14)}$$

where C is a constant quantity taking unit value.

2.7 General permutation theorem

Theorem 7: The number of permutations in a permutation space of N different components taken V at a time where there occurred Y particular components taken from U $\leq$ V particular components (limited size of components) that occurred in the parent component assembly of N components denoted by $P\begin{pmatrix} N & U \\ V & Y \end{pmatrix}$ is

$$P\begin{pmatrix} N & U \\ V & Y \end{pmatrix} = V!\,C\binom{U}{Y}\,C\binom{N-U}{V-Y} \quad\text{------------ (15)}$$

$$\text{where},\ N \geq V$$
$$U \leq V$$
$$k \leq Y \leq U$$
$$0 \leq k = U-N+V$$

Proof: Let the permutation space is of N components taken V at a time. We can select Y components out of U components without restriction is $C\binom{U}{Y}$. Now for each combination we can select $C\binom{N-U}{V-Y}$ combinations. Again for the total combinations each combination can be arranged in V! ways. Thus finally we get the total number of permutations is

$$V!\,C\binom{U}{Y}\,C\binom{N-U}{V-Y}.$$

Example 6: Let the parent component assembly M = (A, B, C, D, E). Find the number of permutations of the following and then write the permutations.

(i) $P\begin{pmatrix} 5 & 3 \\ 3 & 3 \end{pmatrix}$, (ii) $P\begin{pmatrix} 5 & 3 \\ 3 & 1 \end{pmatrix}$, (iii) $P\begin{pmatrix} 5 & 2 \\ 3 & 2 \end{pmatrix}$, (iv) $P\begin{pmatrix} 5 & 2 \\ 3 & 1 \end{pmatrix}$,

(v) $P\begin{pmatrix} 5 & 2 \\ 3 & 0 \end{pmatrix}$.

Solution:

(i) $P\begin{pmatrix} 5 & 3 \\ 3 & 3 \end{pmatrix} = 3! \; C\begin{pmatrix} 3 \\ 3 \end{pmatrix} C\begin{pmatrix} 5-3 \\ 3-3 \end{pmatrix} = 6 \times 1 \times 1 = 6.$

Now the permutations are (A, B, C), (A, C, B), (B, A, C), (B, C, A), (C, A, B), (C, B, A).

(ii) $P\begin{pmatrix} 5 & 3 \\ 3 & 1 \end{pmatrix} = 3! \; C\begin{pmatrix} 3 \\ 1 \end{pmatrix} C\begin{pmatrix} 5-3 \\ 3-1 \end{pmatrix} = 6 \times 3 \times 1 = 18.$

Now the permutations are (A, D, E), (A, E, D), (D, A, E), (D, E, A), (E, A, D), (E, D, A), (B, D, E), (B, E, D), (D, B, E), (D, E, B), (E, B, D), (E, D, B), (C, D, E), (C, E, D), (D, C, E), (D, E, C), (E, C, D), (E, D, C).

(iii) $P\begin{pmatrix} 5 & 2 \\ 3 & 2 \end{pmatrix} = 3! \; C\begin{pmatrix} 2 \\ 2 \end{pmatrix} C\begin{pmatrix} 5-2 \\ 3-2 \end{pmatrix} = 6 \times 1 \times 3 = 18.$

Now the permutations are (A, B, C), (A, C, B), (B, A, C), (B, C, A), (C, A, B), (C, B, A), (A, B, D), (A, D, B), (B, A, D), (B, D, A), (D, A, B), (D, B, A), (A, B, E), (A, E, B), (B, A, E), (B, E, A), (E, A, B), (E, B, A).

(iv) $P\begin{pmatrix} 5 & 2 \\ 3 & 1 \end{pmatrix} = 3! \; C\begin{pmatrix} 2 \\ 1 \end{pmatrix} C\begin{pmatrix} 5-2 \\ 3-1 \end{pmatrix} = 6 \times 2 \times 3 = 36.$

Now the permutations are (A, C, D), (A, D, C), (C, A, D), (C, D, A), (D, A, C), (D, C, A), (A, C, E), (A, E, C), (C, A, E), (C, E, A), (E, A, C), (E, C, A), (A, D, E), (A, E, D), (D, A, E), (D, E, A), (E, A, D), (E, D, A), (B, C, D), (B, D, C), (C, B, D), (C, D, B), (D, B, C), (D, C, B), (B, C, E), (B, E, C), (C, B, E), (C, E, B), (E, B, C), (E, C, B), (B, D, E), (B, E, D), (D, B, E), (D, E, B), (E, B, D), (E, D, B).

(v) $P\begin{pmatrix} 5 & 2 \\ 3 & 0 \end{pmatrix} = 3! \; C\begin{pmatrix} 2 \\ 0 \end{pmatrix} C\begin{pmatrix} 5-2 \\ 3-0 \end{pmatrix} = 6 \times 1 \times 1 = 6.$

Now the permutations are (C, D, E), (C, E, D), (D, C, E), (D, E, C), (E, C, D), (E, D, C).

2.8 Permutation distribution

Theorem 8: A random variable Y is said to follow permutation distribution if it assumes only non-negative values and its probability mass function is given by

$$P(Y) = P(Y; N, U, V) = \frac{V!\, C\binom{U}{Y} C\binom{N-U}{V-Y}}{P\binom{N}{V}} \;;\; Y = k,\, k+1,\, k+2,\, \ldots\ldots,\, U$$

$$0 \le k = U - N + V$$

$$= 0;\ \text{otherwise}. \qquad\qquad\text{————} \quad (16)$$

The three independent finite constants N, U and V are known as the parameters of this distribution. Permutation distribution is a discrete distribution as Y can take only the non-negative values under the interval $U - N + V \le Y \le U$. Any variable which follows permutation distribution is known as permutation variate and denoted by the symbol $Y \sim P(N, U, V)$.

2.8.1 Moments

The first four moments about origin of permutation distribution are

$$\mu_1' = E(Y) = \frac{UV}{N}.$$

$$\mu_2' = E(Y^2) = \frac{UV(U-1)(V-1)}{N(N-1)} + \frac{UV}{N}.$$

$$\mu_3' = E(Y^3) = \frac{UV(U-1)(U-2)(V-1)(V-2)}{N(N-1)(N-2)} + \frac{3UV(U-1)(V-1)}{N(N-1)} + \frac{UV}{N}.$$

$$\mu_4' = E(Y^4)$$
$$= \frac{U(U-1)(U-2)(U-3)V(V-1)(V-2)(V-3)}{N(N-1)(N-2)(N-3)} + \frac{6U(U-1)(U-2)V(V-1)(V-2)}{N(N-1)(N-2)}$$
$$+ \frac{7U(U-1)V(V-1)}{N(N-1)} + \frac{UV}{N}.$$

Now variance $\mu_2 = \mu_2' - \mu_1'^2 = \dfrac{UV(U-1)(V-1)}{N(N-1)} + \dfrac{UV}{N} - \dfrac{U^2V^2}{N^2}$

$$= \frac{UV(N-U)(N-V)}{N^2(N-1)}.$$

2.9 Formation space

A formation space is a set of all possible formations (outcomes) of an experiment from a parent assembly A where the outcomes do not take order of the components into account. Suppose a formation space contains T possible outcomes then the formations space denoted by $F\begin{Bmatrix} A \\ V \end{Bmatrix}$ is

$$F\begin{Bmatrix} A \\ V \end{Bmatrix} = \{F_1, F_2, F_3, \cdots, F_t, \cdots, F_T\} \quad\text{————————} \quad (17)$$

$$\text{where}, \quad V < M, \quad V = M, \quad V > M$$

$$M = \text{Parent components number.}$$

Example 7: Set a formation space of the experiment "a coin tossed 3 times ".

Solution: We have given A = (H, T), V = 3 and M = 2

Thus the formation space is

$$F\begin{Bmatrix} H, T \\ 3 \end{Bmatrix} = \{HHH, \; HHT, HTT, TTT\}.$$

Theorem 9: The number of formations of M sided V dice experiment denoted by $F\binom{M}{V}$ is

$$F\binom{M}{V} = C\binom{M + V - 1}{V} \quad\text{————————} \quad (18)$$

$$\text{where}, \quad V < M$$

$$V = M$$

$$V > M$$

Proof: Let we have V place to be filled with M different components. The M components say arrange in an ordered parent assembly as first component, second component, third component etc. As our definition in a formation a component has the opportunity of repetitions and no opportunity to take order into consideration. Thus it is enough to record the selections which arise from a particular direction. Now we are to have the first place can be filled with one of the M components i.e., k_1 states an index that indicates a component of the O. P. A. taken by first place of a formation takes the interval

$$1 \le k_1 \le M \quad\text{————————} \quad (19)$$

Thus the number of events of formations characterizing first components denoted by S_1 is the number of indexes held by the interval (19) i.e.,

$$S_1 = M - 1 + 1 = M$$

As repetitions of a component may be allowed it would be possible to use the component used in first place in second place where in a combination it is not possible subject to the condition the given components are all district. Thus the second place can be filled with one of the $(M- k_1 + 1)$ components that begins from k_1^{th} component and ends in M^{th} component of the O. P. A. i.e., k_2 takes the interval

$$k_1 \leq k_2 \leq M \qquad\qquad \text{——————————} \qquad (20)$$

obviously when

$k_1 = 1$	then $1 \leq k_2 \leq M$	; M
$k_1 = 2$	then $2 \leq k_2 \leq M$	; M−1
$k_1 = 3$	then $3 \leq k_2 \leq M$	; M−2
$\vdots$		
$k_1 = M$	then $M \leq k_2 \leq M$	; 1

The numbers after semicolons give the number of indexes held by the corresponding intervals. Thus the number of events of formations characterizing second components is the sum of numbers of indexes held by the interval (20) i.e.,

$$S_2 = M + (M-1) + (M-2) + \ldots\ldots + 1 = \frac{M(M+1)}{2!}$$

Now the third place can be filled with one of the $(M - k_2 + 1)$ components that begins from k_2^{th} component and ends in M^{th} component of the O. P. A. i.e., k_3 takes the interval

$$k_2 \leq k_3 \leq M \qquad\qquad \text{——————————} \qquad (21)$$

Clearly when

$k_1 = 1$ and $k_2 = 1$ then	$1 \leq k_3 \leq M$	; M
$k_1 = 1$ and $k_2 = 2$ then	$2 \leq k_3 \leq M$	; M−1
$k_1 = 1$ and $k_2 = 3$ then	$3 \leq k_3 \leq M$	; M−2
$\vdots$		
$k_1 = 1$ and $k_2 = M$ then	$M \leq k_3 \leq M$	; 1
$k_1 = 2$ and $k_2 = 2$ then	$2 \leq k_3 \leq M$	; (M−1)
$k_1 = 2$ and $k_2 = 3$ then	$3 \leq k_3 \leq M$	; (M−2)
$k_1 = 2$ and $k_2 = 4$ then	$4 \leq k_3 \leq M$	; (M−3)
$\vdots$		
$k_1 = 2$ and $k_2 = M$ then	$M \leq k_3 \leq M$	; 1
$\vdots$		

$k_1 = M$ and $k_2 = M$ then $M \le k_3 \le M$ $\qquad$; 1

The numbers after semicolons give the numbers of indexes held by the corresponding intervals. Thus the number of events of formations characterizing third components is the sum of numbers of indexes held by the interval (21) i.e.,

$$S_3 = M + (M-1) + \ldots\ldots + 1 + (M-1) + (M-2) + \ldots\ldots + 1 + \ldots\ldots + 1$$
$$= \frac{M(M+1)(M+3)}{3!}$$

Continuing this process we get the vth place fill with one of the $(M - k_{V-1} + 1)$ components that begins from $k_{V-1}{}^{th}$ component and ends in M^{th} component of the O. P. A. i.e., k_V takes the interval

$$k_{V-1} \le k_V \le M \qquad\qquad \text{————————} \qquad (22)$$

Thus the number of events of formations characterizing V^{th} components is the sum of numbers of indexes held by the interval (22) i.e.,

$$S_V = \frac{M(M+1)(M+2)\ldots(M+V-1)}{V!} = C\binom{M+V-1}{V} \qquad \text{————————} \qquad (23)$$

$$\text{(on simplification)}$$

Since all the places of formations are already characterized thus the expression (23) gives the number of formations of M sided V dice experiment.

2.10 Formation event

It is a special kind of subset of a formation space where the formation members are to be have same first components, same second components, same third components and so on same v^{th} components. Suppose the formation members

$$F_1 = (F_{11}, F_{12}, F_{13}, \ldots, F_{1v}, \ldots, F_{1V})$$
$$F_2 = (F_{21}, F_{22}, F_{23}, \ldots, F_{2v}, \ldots, F_{2V})$$
$$F_3 = (F_{31}, F_{32}, F_{33}, \ldots, F_{3v}, \ldots, F_{3V})$$
$$\vdots$$
$$F_t = (F_{t1}, F_{t2}, F_{t3}, \ldots, F_{tv}, \ldots, F_{tV})$$

$$\text{where, } F_{11} = F_{21} = F_{31} = . \ldots. = F_{t1}$$
$$F_{12} = F_{22} = F_{32} = . \ldots. = F_{t2}$$
$$F_{13} = F_{23} = F_{33} = . \ldots. = F_{t3}$$
$$\vdots$$

14

$$F_{1v} = F_{2v} = F_{3v} = \ldots\ldots = F_{tv}$$

Then the formation event denoted by $F\left\{{A \atop V/^vA}\right\}$ consists of $F_1, F_2, F_3, \cdots, F_t$

i.e.,

$$F\left\{{A \atop V/^vA}\right\} = \{F_1, F_2, F_3, \cdots, F_t\} \hspace{2cm} (24)$$

Here A is a parent assembly contained M components, V is the number of components occurred in a formation member and ^vA is an identified component assembly containing v identified components. The formation event contains any number of formation members and starts with any formation member of $F\left\{{A \atop V}\right\}$ continues the condition supports.

Example 8: Find the formation events of the example 7 where identified components are (i) $^1A = (H)$, $^2A = (H, H)$ and $^3A = (H, H, H)$.

Solution: The formation events are

(i) $F\left\{{(H, T) \atop 3/(H)}\right\} = \{(H, H, H), (H, H, T), (H, T, T)\}$

(ii) $F\left\{{(H, T) \atop 3/(H, H)}\right\} = \{(H, H, H), (H, H, T)\}$

(iii) $F\left\{{(H, T) \atop 3/(H, H, H)}\right\} = \{(H, H, H)\}$.

Theorem 10: The number of formations of M sided V dice experiment under the partition $P_{tQ/q} = \left(P_{t1} + P_{t2} + P_{t3} + \cdots + P_{tq} + \cdots + P_{tQ}\right)$ where the components of the partitions indicate runs or kinds with their associate lengths or lots respectively and whose first q components are identified, denoted by $F\left({M \atop P_{tQ/q}}\right)$ is

$$F\left({M \atop P_{tQ/q}}\right) = \frac{(Q-q)!\, C\binom{M}{Q}}{Q_{tg}!\, Q_{t(g+1)}!\, Q_{t(g+2)}!\, \ldots\ldots Q_{th}!} \hspace{2cm} (25)$$

where , $Q_{tg} + Q_{t(g+1)} + Q_{t(g+2)} + \cdots + Q_{th} = Q-q$

$$Q_{t1} + Q_{t2} + Q_{t3} + \cdots + Q_{t(g-1)} = q$$

and Q_{tg} is the number of components of the partitions alike of a gth kind and so on for $Q_{t(g+1)}, Q_{t(g+2)}, \ldots, Q_{th}$. It is not astonished to you that a gth

kind of component may be alike that of the earlier $(g-1)^{th}$ kind of the partition.

2.11 General formation theorem

Theorem 11: The number of formations in a formation space of M sided V dice experiment where there occurred X particular components taken from U; $M \geq U$ particular components (limited size of components) denoted by $F\begin{pmatrix} M & U \\ V & X \end{pmatrix}$ can be written as

$$F\begin{pmatrix} M & U \\ V & X \end{pmatrix} = C\begin{pmatrix} U \\ X \end{pmatrix} C\begin{pmatrix} M - U + V - 1 \\ V - X \end{pmatrix} \qquad\qquad (26)$$

$$\text{where, } X = 0, 1, 2, \ldots\ldots, \min (M, U, V)$$
$$M \geq U$$

Proof: Let M sided V dice are tossed. We have X particular components taken from U ; $M \geq U$. We know it is enough to count the formations which are ordered or placewise small numbered side to big numbered side. Suppose V places are to be filled. First X places can be filled by U components in $C\begin{pmatrix} U \\ X \end{pmatrix}$ ways. After this is done there are (V−X) places to be filled from (M−U+X) different components. This is performed in $F\begin{pmatrix} M - U + X \\ V - X \end{pmatrix}$ ways as the outcomes are formations. Thus we get

$$F\begin{pmatrix} M & U \\ V & X \end{pmatrix} = C\begin{pmatrix} U \\ X \end{pmatrix} F\begin{pmatrix} M - U + X \\ V - X \end{pmatrix} = C\begin{pmatrix} U \\ X \end{pmatrix} C\begin{pmatrix} M - U + V - 1 \\ V - X \end{pmatrix}$$

Hence the proof.

Example 9: Find the number of formations of the following and the write the formations.

(i) $F\begin{pmatrix} 6 & 3 \\ 4 & 0 \end{pmatrix}$, (ii) $F\begin{pmatrix} 6 & 3 \\ 4 & 3 \end{pmatrix}$, (iii) $F\begin{pmatrix} 4 & 3 \\ 2 & 0 \end{pmatrix}$, (iv) $F\begin{pmatrix} 4 & 3 \\ 2 & 1 \end{pmatrix}$, (v) $F\begin{pmatrix} 4 & 3 \\ 2 & 2 \end{pmatrix}$,
(vi) $F\begin{pmatrix} 3 & 3 \\ 4 & 0 \end{pmatrix}$, (vii) $F\begin{pmatrix} 3 & 3 \\ 4 & 1 \end{pmatrix}$, (viii) $F\begin{pmatrix} 3 & 3 \\ 4 & 2 \end{pmatrix}$, (ix) $F\begin{pmatrix} 3 & 3 \\ 4 & 3 \end{pmatrix}$.

Solutions: Let the parent component assembly A = $(S_1, S_2, S_3, S_4, S_5, S_6)$ and the components of limited size is (S_1, S_2, S_3). Then we get for (i), (ii),(iii) and (iv) the following

(i) $F\begin{pmatrix} 6 & 3 \\ 4 & 0 \end{pmatrix} = C\begin{pmatrix} 3 \\ 0 \end{pmatrix} C\begin{pmatrix} 6 - 3 + 4 - 1 \\ 4 - 0 \end{pmatrix} = C\begin{pmatrix} 3 \\ 0 \end{pmatrix} C\begin{pmatrix} 6 \\ 4 \end{pmatrix} = 1 \times 15 = 15.$

Now the formations are

1. (S_4, S_4, S_4, S_4), 2. (S_4, S_4, S_4, S_5), 3. (S_4, S_4, S_4, S_6), 4. (S_4, S_4, S_5, S_5), 5. (S_4, S_4, S_5, S_6), 6. (S_4, S_4, S_6, S_6), 7. (S_4, S_5, S_5, S_5), 8. (S_4, S_5, S_5, S_6), 9. (S_4, S_5, S_6, S_6), 10. (S_4, S_6, S_6, S_6), 11. (S_5, S_5, S_5, S_5), 12. (S_5, S_5, S_5, S_6), 13. (S_5, S_5, S_6, S_6), 14. (S_5, S_6, S_6, S_6), 15. $(S_6, S_6, S_6. S_6)$.

(ii) $F\begin{pmatrix} 6 & 3 \\ 4 & 3 \end{pmatrix} = C\binom{3}{3}C\binom{6-3+4-1}{4-3} = C\binom{3}{3}C\binom{6}{1} = 1 \times 6 = 6.$

Now the formations are 1. (S_1, S_1, S_2, S_3), 2. (S_1, S_2, S_2, S_3), 3. (S_1, S_2, S_3, S_3), 4. (S_1, S_2, S_3, S_4), 5. (S_1, S_2, S_3, S_5) and 6. (S_1, S_2, S_3, S_6).

Again let the parent component assembly $A = (S_1, S_2, S_3, S_4)$ and the components of limited size is (S_1, S_2, S_3). Then we get for (v), (vi) and (vii) the following

(iii) $F\begin{pmatrix} 4 & 3 \\ 2 & 0 \end{pmatrix} = C\binom{3}{0}C\binom{4-3+2-1}{2-0} = C\binom{3}{0}C\binom{2}{2} = 1 \times 1 = 1.$

Now the formation is (S_4, S_4).

(iv) $F\begin{pmatrix} 4 & 3 \\ 2 & 1 \end{pmatrix} = C\binom{3}{1}C\binom{4-3+2-1}{2-1} = C\binom{3}{1}C\binom{2}{1} = 3 \times 2 = 6.$

Now the formations are 1. (S_1, S_1), 2. (S_1, S_4), 3. (S_2, S_2), 4. (S_2, S_4), 5. (S_3, S_3), 6. (S_3, S_4).

(v) $F\begin{pmatrix} 4 & 3 \\ 2 & 2 \end{pmatrix} = C\binom{3}{2}C\binom{4-3+2-1}{2-2} = C\binom{3}{2}C\binom{2}{0} = 3 \times 1 = 3.$

Now the formations are 1. (S_1, S_2), 2. (S_1, S_3) and 3. (S_2, S_3).

Now let the parent component assembly $A = (S_1, S_2, S_3)$ and the components of limited size is (S_1, S_2, S_3). Then we get for (viii), (ix), (x) and (xi) the following

(vi) $F\begin{pmatrix} 3 & 3 \\ 4 & 0 \end{pmatrix} = C\binom{3}{0}C\binom{3-3+4-1}{4-0} = C\binom{3}{0}C\binom{3}{4} = 1 \times 0 = 0$

(vii) $F\begin{pmatrix} 3 & 3 \\ 4 & 1 \end{pmatrix} = C\binom{3}{1}C\binom{3-3+4-1}{4-1} = C\binom{3}{1}C\binom{3}{3} = 3 \times 1 = 3.$

Now the formations are 1. (S_1, S_1, S_1, S_1), 2. (S_2, S_2, S_2, S_2) and 3. (S_3, S_3, S_3, S_3).

(viii) $F\begin{pmatrix} 3 & 3 \\ 4 & 2 \end{pmatrix} = C\binom{3}{2}C\binom{3-3+4-1}{4-2} = C\binom{3}{2}C\binom{3}{2} = 3 \times 3 = 9.$

Now the formations are 1. (S_1, S_1, S_1, S_2), 2. (S_1, S_1, S_1, S_3), 3. (S_1, S_1, S_2, S_2), 4. (S_1, S_1, S_3, S_3), 5. (S_1, S_2, S_2, S_2), 6. (S_1, S_3, S_3, S_3), 7. (S_2, S_2, S_2, S_3), 8. (S_2, S_2, S_3, S_3), 9. (S_2, S_3, S_3, S_3).

(ix) $F\begin{pmatrix} 3 & 3 \\ 4 & 3 \end{pmatrix} = C\binom{3}{3}C\binom{3-3+4-1}{4-3} = C\binom{3}{3}C\binom{3}{1} = 1 \times 3 = 3.$

Now the formations are 1. (S_1, S_1, S_2, S_3), 2. (S_1, S_2, S_2, S_3) and 3. (S_1, S_2, S_3, S_3).

2.12 Formation distribution

Theorem 12: A random variable X is said to follow formation distribution if it assumes only non negative values and its probability mass function is given by

$$P(X) = f(X; M, U, V) = \frac{c\binom{U}{X} c\binom{M-U+V-1}{V-X}}{F\binom{M}{V}} \; ;$$

$$X = 0, 1, 2, \ldots, \min(M, U, V), \; M \geq U$$

$$= 0 \; ; \text{ others wise} \quad \underline{\hspace{2cm}} \quad (27)$$

The three independent finite constants M, U and V are known as the parameters of this distribution. Formation distribution is a discrete distribution and X can take the values 0, 1, 2, ..., min(M, U, V). A variable which follows formation distribution is known as formation variate. The formation variate is denoted by the symbol $X \sim f(M, U, V)$.

2.12.1 Moments
The first four moments about origin of formation distribution are obtained as follows:

$$\mu_1' = \frac{UV}{M+V-1}.$$

$$\mu_2' = \frac{UV(U-1)(V-1)}{(M+V-1)(M+V-2)} + \frac{UV}{(M+V-1)}.$$

$$\mu_3' = \frac{U(U-1)(U-2)V(V-1)(V-2)}{(M+V-1)(M+V-2)(M+V-3)} + \frac{3U(U-1)V(V-1)}{(M+V-1)(M+V-2)} + \frac{UV}{(M+V-1)}.$$

$$\mu_4' = \frac{U(U-1)(U-2)(U-3)V(V-1)(V-2)(V-3)}{(M+V-1)(M+V-2)(M+V-3)(M+V-4)} + \frac{6U(U-1)(U-2)V(V-1)(V-2)}{(M+V-1)(M+V-2)(M+V-3)}$$

$$+ \frac{7U(U-1)V(V-1)}{(M+V-1)(M+V-2)} + \frac{UV}{(M+V-1)}.$$

Now variance $\mu_2 = \mu_2' - \mu_1'^2 = \frac{UV(M-1)(M-U+V-1)}{(M+V-1)^2(M+V-2)}.$

2.13 Homogenation space

A homogenation space is a set of all possible homogenations (outcomes) of an experiment from a parent assembly A where the outcomes take order of the components into account. Suppose a homogenation space contains T possible outcomes then the homogenation space denoted by $H\{^A_V\}$ is

$$H\{^A_V\} = \{H_1, H_2, H_3, \ldots\ldots H_t, \ldots\ldots H_T\} \quad\text{————(28)}$$

where , $V < M$
$$V = M$$
$$V > M$$
$$M = \text{Parent components number.}$$

Example 10: Set a homogenation space of the experiment "a 3 sided dice tossed 3 times"

Solution: We have given $A = \{S_1, S_2, S_3\}$, $V = 3$ and $M = 3$

Here $S_1 = $ first side, $S_2 = $ second side and $S_3 = $ third side.

Thus the homogenation space is

$$H\left\{\frac{(S_1, S_2, S_3)}{3}\right\} = \{(S_1\ S_1\ S_1),\ (S_1\ S_1\ S_2),\ (S_1\ S_1\ S_3),\ (S_1\ S_2\ S_1),\ (S_1\ S_2\ S_2),$$

$(S_1\ S_2\ S_3),\ (S_1\ S_3\ S_1),\ (S_1\ S_3\ S_2),\ (S_1\ S_3\ S_3),\ (S_2\ S_1\ S_1),\ (S_2\ S_1\ S_2),\ (S_2\ S_1\ S_3),$ $(S_2\ S_2\ S_1),\ (S_2\ S_2\ S_2),\ (S_2\ S_2\ S_3),\ (S_2\ S_3\ S_1),\ (S_2\ S_3\ S_2),\ (S_2\ S_3\ S_3),\ (S_3\ S_1\ S_1),$ $(S_3\ S_1\ S_2),\ (S_3\ S_1\ S_3),\ (S_3\ S_2\ S_1),\ (S_3\ S_2\ S_2),\ (S_3\ S_2\ S_3),\ (S_3\ S_3\ S_1),\ (S_3\ S_3\ S_2),$ $(S_3\ S_3\ S_3)\}$.

Theorem 13: The number of homogenations of M sided V dice experiment denoted by $H\binom{M}{V}$ is

$$H\binom{M}{V} = M^V \quad\text{————— (29)}$$

where, $M < V$, $M = V$, $M > V$

Proof: The theorem is an elementary one. Suppose M sided one die is tossed. Then the number of possible outcomes is M. Again M sided two dice are tossed. Then each of M sides of the first die is associated with M sides of the first die, that is the number of possible outcomes is $M.M. = M^2$. If M sided three dice are tossed then each of M sides of third die is associated with the possible outcomes that are associated with the first two dice. Thus the number of possible outcomes is $M^2.M = M^3$. The pattern shows that the number of homogenations of M sided V dice experiment is M^V.

Example 11: Find the number of homogenations of 6 sided 10 dice experiment.

Solution : $H\begin{pmatrix} 6 \\ 10 \end{pmatrix} = 6^{10} = 60466176.$

Theorem 14: The number of homogenations of M sided V dice experiment under the partition $P_t = (P_{t1} + P_{t2} + P_{t3} + \cdots + P_{tq} + \cdots + P_{tQ})$ where the components of the partition indicate runs and their associate lengths denoted by $H\begin{pmatrix} M \\ P_t \end{pmatrix}_{\to r.l}$ is

$$H\begin{pmatrix} M \\ P_t \end{pmatrix}_{\to r.l} = \frac{Q! \, M(M-1)^{Q-1}}{Q_{t1}! \, Q_{t2}! \, Q_{t3}! \dots Q_{th}!} \qquad\qquad (30)$$

$$\text{where}, \; 1 \le h \le Q$$

Example 12: How many homogenations are there of 6 sided 10 dice experiment having

(i) 4 runs, three are of length 2 and one is of length 4

(ii) 4 runs, one is of length 1 and three are of length 3

Solution: (i) This problem is nothing but the number of homogenations under the partition $P_8 = (2+2+2+4)$ of 10 where the components indicate runs and their associate lengths. Now we have given

$M = 6,\; Q = 4,\; Q_{8.1} = 3,\; Q_{8.2} = 1$ and $P_8 = (2+2+2+4)$.

Thus from the theorem 4.4 we get the desired number of homogenations is

$$H\begin{pmatrix} 6 \\ P_8 \end{pmatrix}_{\to r.l} = \frac{4! \, 6.5^3}{3! \, 1!} = 3000.$$

(ii) This is the number of homogenations under the partition $P_7 = (1+3+3+3)$ where the components indicate runs and their associate lengths. Now we have given

$M = 6,\; Q = 4,\; Q_{7.1} = 1,\; Q_{7.2} = 3$ and $P_7 = (1+3+3+3)$.

Thus the desired number of homogenations is

$$H\begin{pmatrix} 6 \\ P_7 \end{pmatrix}_{\to r.l} = \frac{4! \, 6.5^3}{1! \, 3!} = 3000.$$

Theorem 15: The number of homogenations of M sided V dice experiment under the partition $P_t = (P_{t1} + P_{t2} + P_{t3} + \cdots + P_{tQ})$ where the components of the partition indicate kinds and their associate lots, denoted

by $H\binom{M}{P_t}_{\to k.l}$ is

$$H\binom{M}{P_t}_{\to k.l} = \frac{V!\,P\binom{M}{Q}}{P_{t1}!\,P_{t2}!\,P_{t3}!\,......P_{tQ}!\,Q_{t1}!\,Q_{t2}!\,Q_{t3}!\,......Q_{th}!} \qquad\qquad (31)$$

$$\text{where}, \; 1 \leq Q \leq M \leq V$$
$$1 \leq Q \leq V < M$$
$$1 \leq h \leq Q$$

Example 13: How many homogenations are there of 6 sided die tossing 6 times having (i) 3 kinds ? and (ii) 3 kinds ; each of lot 2.

Solution: We have given M = 6, V = 6 and Q = 3.

(i) Now we find the partitions of 6 occuring 3 components. Since they are $P_1 = (1 + 1 + 4)$, $P_2 = (1 + 2 + 3)$ and $P_3 = (2 + 2 + 2)$.

Thus the number of homogenations having 3 kinds is

$$H\binom{6}{P_1}_{\to k.l} + H\binom{6}{P_2}_{\to k.l} + H\binom{6}{P_3}_{\to k.l} = \frac{6!\,P\binom{6}{3}}{4!\,2!} + \frac{6!\,P\binom{6}{3}}{2!\,3!} + \frac{6!\,P\binom{6}{3}}{2!\,2!\,2!\,3!}$$

$$= 1800 + 7200 + 1800 = 10800.$$

(ii) The number of homogenations having 3 kinds each of lot 2 is given by

$$H\binom{6}{P_3}_{\to k.l} = \frac{6!\,P\binom{6}{3}}{2!\,2!\,2!\,3!} = 1800.$$

2.14 Homogenation event

It is a special kind of subset of a homogenation space where the homogenation members are to be have same first components, same second components, same third components and so on same v^{th} components. Suppose the homogenation members are

$$H_1 = (H_{11}, H_{12}, H_{13}, ..., H_{1v}, ..., H_{1V})$$
$$H_2 = (H_{21}, H_{22}, H_{23}, ..., H_{2v}, ..., H_{2V})$$
$$H_3 = (H_{31}, H_{32}, H_{33}, ..., H_{3v}, ..., H_{3V})$$
$$\vdots$$
$$H_t = (H_{t1}, H_{t2}, H_{t3}, ..., H_{tv}, ..., H_{tV})$$

$$\text{where}, \; H_{11} = H_{21} = H_{31} = ... = H_{t1}$$
$$H_{12} = H_{22} = H_{32} = ... = H_{t2}$$

$$H_{13} = H_{23} = H_{33} = \dots = H_{t3}$$
$$\vdots$$
$$H_{1v} = H_{2v} = H_{3v} = \dots = H_{tv}$$

Then the homogenation event denoted by $H\left\{{A \atop V/{}^vA}\right\}$ consists of $H_1, H_2, H_3, \cdots, H_t$ i.e.,

$$H\left\{{A \atop V/{}^vA}\right\} = \{H_1, H_2, H_3, \cdots, H_t\} \quad\text{———————} \quad (32)$$

Here A is a parent assembly containing M components, V is the number of components occurred in a homogenation member and ${}^vA$ is an identified component assembly containing v identified components.

Example 14: Find the homogenation events of the example 10 where the identified components are (i) ${}^1A = (S_2)$, (ii) ${}^2A = (S_2 S_1)$ and (iii) ${}^2A = (S_3 S_2)$

Solution: The homogenation events are

(i) $H\left\{{(S_1, S_2, S_3) \atop 3/(S_2)}\right\} = \{(S_2 S_1 S_1), (S_2 S_1 S_2), (S_2 S_1 S_3), (S_2 S_2 S_1), (S_2 S_2 S_2),$ $(S_2 S_2 S_3), (S_2 S_3 S_1), (S_2 S_3 S_2), (S_2 S_3 S_3)\}$.

(ii) $H\left\{{(S_1, S_2, S_3) \atop 3/(S_2, S_1)}\right\} = \{(S_2 S_1 S_1), (S_2 S_1 S_2), (S_2 S_1 S_3)\}$.

(iii) $H\left\{{(S_1, S_2, S_3) \atop 3/(S_3, S_2)}\right\} = \{(S_3 S_2 S_1), (S_3 S_2 S_2), (S_3 S_2 S_3)\}$.

Theorem 16: The number of homogenations of M sided V dice experiment under the partitions $P_{tQ/q} = \left(P_{t1} + P_{t2} + P_{t3} + \cdots + P_{tq} + \cdots + P_{tQ}\right)$ where the components of the partition indicate runs and their associates lengths and whose first q components are identified denoted by $H\left({M \atop P_{tQ/q}}\right)_{\to r.l}$ is

$$H\left({M \atop P_{tQ/q}}\right)_{\to r.l} = \frac{(Q-q)! \, M(M-1)^{Q-1}}{Q_{tg}! \, Q_{t(g+1)}! \cdots Q_{th}!} \quad\text{———————}\quad (33)$$

$$\text{where, } Q_{t1} + Q_{t2} + Q_{t3} + \cdots + Q_{t(g-1)} = q$$
$$Q_{tg} + Q_{t(g+1)} + Q_{t(g+2)} + \cdots + Q_{th} = Q - q$$

and Q_{tg} is the number of components of the partitions alike of a gth kind and so on for $Q_{t(g+1)}, Q_{t(g+2)}, \dots, Q_{th}$. It is not astonished to you that a gth kind of components in this case may be alike that of the earlier $(g-1)^{th}$ kind of the partition.

Theorem 17: The number of homogenations of M sided V dice experiment under the partition $P_{tQ/q} = \left(P_{t1} + P_{t2} + P_{t3} + \dots + P_{tq} + P_{t(q+1)} + P_{t(q+2)} + \dots + P_{tQ} \right)$ where the components of the partition indicate kinds and their associate lots and whose first q components are identified denoted by $H\binom{M}{P_{tQ/q}}_{\to k.l}$ is

$$H\binom{M}{P_{tQ/q}}_{\to k.l} = \frac{(V - P_{t1} - P_{t2} - P_{t3} - \dots - P_{tq})\,! \; P\binom{M}{Q}}{P_{t(q+1)}\,! \; P_{t(q+2)}\,! \dots P_{tQ}\,! \; Q_{tg}\,! \; Q_{t(g+1)}\,! \dots Q_{th}\,!} \quad\text{——— (34)}$$

$$\text{where, } P_{t(q+1)} + P_{t(q+2)} + \dots + P_{tQ} = V - P_{t1} - P_{t2} - \dots - P_{tq}$$

$$Q_{tg} + Q_{t(g+1)} + \dots + Q_{th} = Q - q$$

$$Q_{t1} + Q_{t2} + \dots + Q_{t(g-1)} = q$$

and Q_{tg} is the number of components of the partition alike of a gth kind and so on for $Q_{t(g+1)}, Q_{t(g+2)}, \dots, Q_{th}$. It is not astonished to you that a gth kind of components in this case may be alike that of the earlier $(g-1)^{th}$ kind of the partition.

2.15 Multinomial expression

Now we state an expression called Multinomial Expression that widely used in the future theorems in this chapter. The multinomial expression denoted by $E_X^V(a)$ called multinomial expression of degree 'V' and range 'X' states as

$$\begin{aligned}
E_X^V(a) = \; & a_1^V + a_2^V + a_3^V + \dots + a_X^V \\
& + a_1^{V-1} a_2 + a_1^{V-1} a_3 + a_1^{V-1} a_4 + \dots + a_1^{V-1} a_X \\
& + a_2^{V-1} a_1 + a_2^{V-1} a_3 + a_2^{V-1} a_4 + \dots + a_2^{V-1} a_X \\
& + \dots + a_{X-1}^{V-1} a_X \\
& + a_1^{V-2} a_2^2 + a_1^{V-2} a_3^2 + a_1^{V-2} a_4^2 + \dots + a_1^{V-2} a_X^2 \\
& + a_2^{V-2} a_1^2 + a_2^{V-2} a_3^2 + a_2^{V-2} a_4^2 + \dots + a_2^{V-2} a_X^2 + \dots + a_{X-1}^{V-2} a_X^2
\end{aligned}$$

23

$$+ a_1^{V-2}a_2a_3 + a_1^{V-2}a_2a_4 + a_1^{V-2}a_2a_5 + \ldots + a_1^{V-2}a_2a_X$$
$$+ a_2^{V-2}a_1a_3 + a_2^{V-2}a_1a_4 + a_2^{V-2}a_1a_5 + \ldots + a_2^{V-2}a_1a_X$$
$$+ \ldots + a_{X-2}^{V-2}a_{X-1}a_X + \ldots$$
$$+ a_1a_2a_3 \ldots a_X \text{ or, } a_1a_2a_3 \ldots a_X^{V-X+1} \text{ or } a_{X-V+1}a_{X-V+2}a_{X-V+3} \cdots a_X.$$

The last term provided as $V = X$ or $V > X$ or, $V < X$.

Now we get the expression as

$$E_X^V(a) = \sum a^V + \sum a_1^{V-1}a_2 + \sum a_1^{V-2}a_2^2 + \sum a_1^{V-2}a_2a_3 + \ldots$$
$$+ \sum a_1a_2a_3 \ldots a_X \text{ or, } \sum a_1a_2a_3 \ldots a_X^{V-X+1} \text{ or, } \sum a_1a_2a_3 \ldots a_V$$

Extending the terms in the right side we get the multinomial expression of degree "V" and range "X" as

$$E_X^V(a) = \sum a^V + \sum a_1^{V-1}a_2 + \sum a_1^{V-2}a_2^2 + \sum a_1^{V-2}a_2a_3 + \sum a_1^{V-3}a_2^3$$
$$+ \sum a_1^{V-3}a_2^2a_3 + \sum a_1^{V-3}a_2a_3a_4 + \ldots$$
$$+ \sum a_1a_2a_3 \ldots a_X \text{ or, } \sum a_1a_2a_3 \ldots a_X^{V-X+1} \text{ or, } \sum a_1a_2a_3 \ldots a_V$$

$$\rule{5cm}{0.4pt} \quad (35)$$

The last term provided as $V = X$ or $V > X$ or $V < X$. There is no need to prove the expression.

Example 15: Set the multinomial expression of degree 3, 4 and 5 of following variables a, b, c and d.

Solution: (i) $E_4^3 = \sum a^3 + \sum a^2b + \sum abc$

$$= a^3 + b^3 + c^3 + d^3 + a^2b + a^2c + a^2d + b^2a + b^2c + b^2d$$
$$+ c^2a + c^2b + c^2d + d^2a + d^2b + d^2c + abc + abd + acd + bcd$$

(ii) $E_4^4 = \sum a^4 + \sum a^3b + \sum a^2b^2 + \sum a^2bc + \sum abcd$

$$= a^4 + b^4 + c^4 + d^4 + a^3b + a^3c + a^3d + b^3a + b^3c + b^3d + c^3a + c^3b + c^3d + d^3a$$
$$+ d^3b + d^3c + a^2b^2 + a^2c^2 + a^2d^2 + b^2c^2 + b^2d^2 + c^2d^2 + a^2bc + a^2bd$$
$$+ a^2cd + b^2ac + b^2ad + b^2cd + c^2ab + c^2ad + c^2bd + d^2ab + d^2ad + d^2bc + abcd$$

(iii) $E_4^5 = \sum a^5 + \sum a^4b + \sum a^3b^2 + \sum a^3bc + \sum a^2b^2c + \sum a^2bcd$

$$= a^5 + b^5 + c^5 + d^5 + a^4b + a^4c + a^4d + b^4a + b^4c + b^4d + c^4a + c^4b + c^4d + d^4a$$
$$+ d^4b + d^4c + a^3b^2 + a^3c^2 + a^3d^2 + b^3a^2 + b^3c^2 + b^3d^2 + c^3a^2 + c^3b^2 + c^3d^2 + d^3a^2$$
$$+ d^3b^2 + d^3c^2 + a^3bc + a^3bd + a^3cd + b^3ac + b^3ad + b^3cd + c^3ab + c^3ad + c^3bd$$
$$+ d^3ab + d^3ac + d^3bc + a^2b^2c + a^2b^2d + a^2c^2b + a^2c^2d + a^2d^2b + a^2d^2c + b^2c^2a$$
$$+ b^2c^2d + b^2d^2a + b^2d^2c + c^2d^2a + c^2d^2b + a^2bcd + ab^2cd + abc^2d + abcd^2.$$

2.16 General homogenation theorem

Theorem 18: The number of homogenations in a homogenation space of M sided V dice experiment where there occurred X particular components taken from U; M ≥ U particular components (limited size of components) denoted by $H\begin{pmatrix} M & U \\ V & X \end{pmatrix}$ can be written as

$$H\begin{pmatrix} M & U \\ V & X \end{pmatrix} = U^{(X)}E_{X+1}^{V-X}(a) \qquad\qquad\text{———(36)}$$

where, $a_1 = M-U$

$\qquad\qquad a_2 = M-U+1$

$\qquad\qquad a_3 = M-U+2$

$\qquad\qquad \vdots$

$\qquad\qquad a_{X+1} = M-U+X$

$\qquad\qquad$ and $X = 0, 1, 2, \ldots\ldots, \min(M, U, V)$; $M \geq U$

Proof: Suppose M sided 1 die is tossed. Then there produced the B space

$1, 2, 3, \ldots\ldots, U, \ldots\ldots, M$

and we get

$$H\begin{pmatrix} M & U \\ 1 & 0 \end{pmatrix} = (M-U) = U^{(0)}E_{0+1}^{1-0}(M-U)$$

$$H\begin{pmatrix} M & U \\ 1 & 1 \end{pmatrix} = U = U^{(1)}E_{0+1}^{1-1}(M-U)$$

Again M sided 2 dice is tossed. Then there produced the B space

$11, 12, 13, \ldots\ldots, 1U, \ldots\ldots, 1M$

$21, 22, 23, \ldots\ldots, 2U, \ldots\ldots, 2M$

$31, 32, 33, \ldots\ldots, 3U, \ldots\ldots, 3M$

$\vdots$

$U1, U2, U3, \ldots\ldots, UU, \ldots\ldots, UM$

$\vdots$

$M1, M2, M3, \ldots\ldots, MU, \ldots\ldots, MM$

and we get

$$H\begin{pmatrix} M & U \\ 2 & 0 \end{pmatrix} = (M-U)^2 = U^{(0)}E_{0+1}^{2-0}(M-U)$$

$$H\begin{pmatrix} M & U \\ 2 & 1 \end{pmatrix} = U(M-U) + (M-U+1)U = U^{(1)}E_{1+1}^{2-1}(M-U) \text{ and}$$

$$H\begin{pmatrix} M & U \\ 2 & 2 \end{pmatrix} = U(U-1) = U^{(2)}E_{2+1}^{2-2}(M-U)$$

After this is done M sided 3 dice is tossed. Then there produce the B space

111, 112, 113, ..., 11U, ..., 11M

121, 122, 123, ..., 12U, ..., 12M

131, 132, 133, ..., 13U, ..., 13M

⋮

1U1, 1U2, 1U3, ..., 1UU, ..., 1UM

⋮

1M1, 1M2, 1M3, ..., 1MU, ..., 1MM

211, 212, 213, ..., 21U, ..., 21M

221, 222, 223, ..., 22U, ..., 22M

231, 232, 233, ..., 23U, ..., 23M

⋮

2U1, 2U2, 2U3, ..., 2UU, ..., 2UM

⋮

2M1, 2M2, 2M3, ..., 2MU, ..., 2MM

311, 312, 313, ..., 31U, ..., 31M

321, 322, 323, ..., 32U, ..., 32M

331, 332, 333, ..., 33U, ..., 33M

⋮

3U1, 3U2, 3U3, ..., 3UU, ..., 3UM

⋮

3M1, 3M2, 3M3, ..., 3MU, ..., 3MM

⋮

U11, U12, U13, ..., U1U, ..., U1M

U21, U22, U23, ..., U2U, ..., U2M

U31, U32, U33, ..., U3U, ..., U3M

⋮

UU1, UU2, UU3, ..., UUU, ..., UUM

⋮

UM1, UM2, UM3, ..., UMU, ..., UMM

⋮

M11, M12, M13, ..., M1U, ..., M1M

M21, M22, M23, ..., M2U, ..., M2M

M31, M32, M33, ..., M3U, ..., M3M

$\vdots$

MU1, MU2, MU3, ..., MUU, ..., MUM

$\vdots$

MM1, MM2, MM3, ..., MMU, ..., MMM

And we get

$$H\begin{pmatrix} M & U \\ 3 & 0 \end{pmatrix} = (M-U)^3 = U^{(0)}E^{3-0}_{0+1}(M-U)$$

$$H\begin{pmatrix} M & U \\ 3 & 1 \end{pmatrix} = U(M-U)^2 + U(M-U)(M-U+1) + U(M-U+1)^2$$

$$= U^{(1)}E^{3-1}_{1+1}(M-U)$$

$$H\begin{pmatrix} M & U \\ 3 & 2 \end{pmatrix} = U(U-1)+(M-U)+U(U-1)(M-U+1)+ U(U-1)(M-U+2)$$

$$= U^{(2)}E^{3-2}_{2+1}(M-U)$$

$$H\begin{pmatrix} M & U \\ 3 & 3 \end{pmatrix} = U(U-1)(U-2) = U^{(3)}E^{3-3}_{3+1}(M-U)$$

Proceeding these ways we get for

$$H\begin{pmatrix} M & U \\ 4 & 0 \end{pmatrix} = U^{(0)}E^{4-0}_{0+1}(M-U)$$

$$H\begin{pmatrix} M & U \\ 4 & 1 \end{pmatrix} = U^{(1)}E^{4-1}_{1+1}(M-U)$$

$$H\begin{pmatrix} M & U \\ 4 & 2 \end{pmatrix} = U^{(2)}E^{4-2}_{2+1}(M-U)$$

$$H\begin{pmatrix} M & U \\ 4 & 3 \end{pmatrix} = U^{(3)}E^{4-3}_{3+1}(M-U)$$

$$H\begin{pmatrix} M & U \\ 4 & 4 \end{pmatrix} = U^{(4)}E^{4-4}_{4+1}(M-U)$$

Continuing these process we get at last

$$H\begin{pmatrix} M & U \\ V & X \end{pmatrix} = U^{(X)}E^{V-X}_{X+1}(M-U)$$

$$\text{where, } X = 0, 1, 2, \ldots, \min(M, U, V)$$

Example 16: Find the number of homogenations of the following

(i) $H\begin{pmatrix} 6 & 3 \\ 4 & 0 \end{pmatrix}$, (ii) $H\begin{pmatrix} 6 & 3 \\ 4 & 1 \end{pmatrix}$, (iii) $H\begin{pmatrix} 6 & 3 \\ 4 & 2 \end{pmatrix}$, (iv) $H\begin{pmatrix} 6 & 3 \\ 4 & 3 \end{pmatrix}$,

(v) $H\begin{pmatrix} 8 & 6 \\ 4 & 0 \end{pmatrix}$, (vi) $H\begin{pmatrix} 8 & 6 \\ 4 & 1 \end{pmatrix}$, (vii) $H\begin{pmatrix} 8 & 6 \\ 4 & 2 \end{pmatrix}$, (viii) $H\begin{pmatrix} 8 & 6 \\ 4 & 3 \end{pmatrix}$,

(ix) $H\begin{pmatrix} 8 & 6 \\ 4 & 4 \end{pmatrix}$, (x) $H\begin{pmatrix} 4 & 4 \\ 5 & 0 \end{pmatrix}$, (xi) $H\begin{pmatrix} 4 & 4 \\ 5 & 1 \end{pmatrix}$, (xii) $H\begin{pmatrix} 4 & 4 \\ 5 & 2 \end{pmatrix}$,

(xiii) $H\begin{pmatrix} 4 & 4 \\ 5 & 3 \end{pmatrix}$, (xiv) $H\begin{pmatrix} 4 & 4 \\ 5 & 4 \end{pmatrix}$.

Solution: Let the parent component assembly $A = (S_1, S_2, S_3, S_4, S_5, S_6)$ and the components of limited size is (S_1, S_2, S_3). Then we get

(i) $H\begin{pmatrix} 6 & 3 \\ 4 & 0 \end{pmatrix} = 3^{(0)}E_{0+1}^{4-0}(6-3) = 1[3^4] = 1 \times 81 = 81$

(ii) $H\begin{pmatrix} 6 & 3 \\ 4 & 1 \end{pmatrix} = 3^{(1)}E_{1+1}^{4-1}(6-3)$

$= 3[3^3 + 4^3 + 3^2 \times 4 + 3 \times 4^2] = 3 \times 175 = 525$

(iii) $H\begin{pmatrix} 6 & 3 \\ 4 & 2 \end{pmatrix} = 3^{(2)}E_{2+1}^{4-2}(6-3)$

$= 3 \times 2[3^2 + 4^2 + 5^2 + 3 \times 4 + 3 \times 5 + 4 \times 5] = 582$

(iv) $H\begin{pmatrix} 6 & 3 \\ 4 & 3 \end{pmatrix} = 3^{(3)}E_{3+1}^{4-3}(6-3)$

$= 3 \times 2 \times 1[3 + 4 + 5 + 6] = 6 \times 18 = 108$

(v) $H\begin{pmatrix} 8 & 6 \\ 4 & 0 \end{pmatrix} = 6^{(0)}E_{0+1}^{4-0}(8-6) = 1[2^4] = 1 \times 16 = 16$

(vi) $H\begin{pmatrix} 8 & 6 \\ 4 & 1 \end{pmatrix} = 6^{(1)}E_{1+1}^{4-1}(8-6)$

$= 6[2^3 + 3^3 + 2^2 \times 3 + 2 \times 3^2] = 6 \times 65 = 390$

(vii) $H\begin{pmatrix} 8 & 6 \\ 4 & 2 \end{pmatrix} = 6^{(2)}E_{2+1}^{4-2}(8-6)$

$= 30[2^2 + 3^2 + 4^2 + 2 \times 3 + 2 \times 4 + 3 \times 4] = 1650$

(viii) $H\begin{pmatrix} 8 & 6 \\ 4 & 3 \end{pmatrix} = 6^{(3)}E_{3+1}^{4-3}(8-6)$

$= 120[2 + 3 + 4 + 5] = 120 \times 14 = 1680$

(ix) $H\begin{pmatrix} 8 & 6 \\ 4 & 4 \end{pmatrix} = 6^{(4)}E_{4+1}^{4-4}(8-6) = 360[2^0] = 360 \times 1 = 360$

(x) $H\begin{pmatrix} 4 & 4 \\ 5 & 0 \end{pmatrix} = 4^{(0)}E_{0+1}^{5-0}(4-4) = 1[0^5] = 1 \times 0 = 0$

(xi) $H\begin{pmatrix} 4 & 4 \\ 5 & 1 \end{pmatrix} = 4^{(1)}E_{1+1}^{5-1}(4-4)$

$= 4[0^4 + 1^4 + 0^3 \times 1 + 0 \times 1^3] = 4 \times 1 = 4$

(xii) $H\begin{pmatrix} 4 & 4 \\ 5 & 2 \end{pmatrix} = 4^{(2)}E_{2+1}^{5-2}(4-4)$

$= 12[0^3 + 1^3 + 2^3 + 0^2 \times 1 + 0^2 \times 2 + 1^2 \times 2$

$+ 0 \times 1^2 + 0 \times 2^2 + 1 \times 2^2 + 0 \times 1 \times 2] = 180$

(xiii) $H\begin{pmatrix} 4 & 4 \\ 5 & 3 \end{pmatrix} = 4^{(3)}E_{3+1}^{5-3}(4-4)$

$= 24[0^2 + 1^2 + 2^2 + 3^3 + 0 \times 1 + 0 \times 2 + 0 \times 3 + 1 \times 2$

$+ 1 \times 3 + 2 \times 3] = 600$

(xiv) $H\begin{pmatrix} 4 & 4 \\ 5 & 4 \end{pmatrix} = 4^{(4)}E_{4+1}^{5-4}(4-4)$

$$= 24[0 + 1 + 2 + 3 + 4] = 24 \times 10 = 240.$$

2.17 Homogenation distribution

Theorem 19: A random variable X is said to follow homogenation distribution if it assumes only non-negative values and its probability mass function is given by

$P(X) = h(X; M, U, V)$

$= \dfrac{U^{(X)}E_{X+1}^{V-X}(M-U)}{H\binom{M}{V}}$; $X = 0, 1, 2, \ldots\ldots, \min(M, U, V)$; $M \geq U$

$= 0$; otherwise   ————— (37)

The three independent finite constants M, U and V are known as the parameters of this distribution. Homogenation distribution is a discrete distribution as X can take only the values 0, 1, 2, ..., min(M, U, V). The variable which follows homogenation distribution is known as homogenation variate. A homogenation variate is denoted by the symbol $X \sim h (M, U, V)$.

3. Conclusions

The four Biswas members are combinations, permutations, formations and homogenations. Combinations and permutations are known to us but formations and homogenations are new ideas. Beside identified and general combination theorems, identified and general permutation theorems are introduced in this paper.

References

1. Deapon Biswas, Paper 2, Assemblies, Bystematics My Classic, 2010 Self published, Chittagong, 2016 Monon Prokashon, Chittagong, Bystematics Vol. I, My Classic, 2018 Scholar's Press EU, ISBN: 987- 620-2-30664-5.

2. Deapon Biswas, Paper 4, B space, Bystematics My Classic, 2010 Self published, Chittagong, 2016 Monon Prokashon, Chittagong, Bystematics Vol. I, My Classic, 2018 Scholar's Press EU, ISBN: 987- 620-2-30664-5.

3. Deapon Biswas, Paper 6, Summation methods, Bystematics My Classic, 2010 Self published, Chittagong, 2016 Monon Prokashon, Chittagong, Bystematics Vol. I, My Classic, 2018 Scholar's Press EU, ISBN: 987- 620-2-30664-5.

4. Deapon Biswas, Paper 7, On the partitions, Bystematics My Classic, 2010 Self published, Chittagong, 2016 Monon Prokashon, Chittagong, Bystematics Vol. I, My Classic, 2018 Scholar's Press EU, ISBN: 987- 620-2-30664-5.

5. Deapon Biswas, Paper 8, Arranged partitions, Bystematics My Classic, 2010 Self published, Chittagong, 2016 Monon Prokashon, Chittagong, Bystematics Vol. I, My Classic, 2018 Scholar's Press EU, ISBN: 987- 620-2-30664-5.

6. Deapon Biswas, Paper 13, On the combinations, Bystematics My Classic, 2010 Self published, Chittagong, 2016 Monon Prokashon, Chittagong, Bystematics Vol. II, My Classic, 2018 Scholar's Press EU, ISBN: 987- 620-2-30960-8.

7. Deapon Biswas, Paper 16, On the permutations, Bystematics My Classic, 2010 Self published, Chittagong, 2016 Monon Prokashon, Chittagong, Bystematics Vol. II, My Classic, 2018 Scholar's Press EU, ISBN: 987- 620-2-30960-8.

8. Deapon Biswas, Paper 18, Formations, Bystematics My Classic, 2010 Self published, Chittagong, 2016 Monon Prokashon, Chittagong, Bystematics Vol. II, My Classic, 2018 Scholar's Press EU, ISBN: 987- 620-2-30960-8.

9. Deapon Biswas, Paper 20, Homogenations, Bystematics My Classic, 2010 Self published, Chittagong, 2016 Monon Prokashon, Chittagong, Bystematics Vol. II, My Classic, 2018 Scholar's Press EU, ISBN: 987- 620-2-30960-8.

YOUR KNOWLEDGE HAS VALUE

- We will publish your bachelor's and master's thesis, essays and papers

- Your own eBook and book - sold worldwide in all relevant shops

- Earn money with each sale

Upload your text at www.GRIN.com
and publish for free